We Belong to Gaia

We Belong to Gaia

JAMES LOVELOCK

PENGUIN BOOKS — GREEN IDEAS

PENGUIN BOOKS

UK | USA | Canada | Ireland | Australia
India | New Zealand | South Africa

Penguin Books is part of the Penguin Random House group of companies
whose addresses can be found at global.penguinrandomhouse.com.

Penguin
Random House
UK

First published in *The Revenge of Gaia* by Penguin Books 2006
This selection published in Penguin Books 2021

003

Copyright © James Lovelock, 2006

Set in 12.5/15pt Dante MT Std
Typeset by Jouve (UK), Milton Keynes
Printed and bound in Great Britain by Clays Ltd, Elcograf S.p.A.

The authorized representative in the EEA is Penguin Random House Ireland,
Morrison Chambers, 32 Nassau Street, Dublin D02 YH68

A CIP catalogue record for this book is available from the British Library

ISBN: 978-0-241-51464-1

www.greenpenguin.co.uk

Contents

What is Gaia?

Hardly anyone, and that included me for the first ten years after the concept was born, seems to know what Gaia is. Most scientists, when they think and talk about the living part of he Earth, call it the biosphere, although strictly speaking the biosphere is no more than the geographical region where life exists, the thin spherical bubble at the Earth's surface. They have unconsciously expanded the definition of the biosphere into something larger than a geographical region but seem vague about where it starts and ends geographically and what it does.

Going outwards from the centre, the Earth is almost entirely made of hot or molten rock and metal. Gaia is a thin spherical shell of matter that surrounds the incandescent interior; it begins where the crustal rocks meet the magma of the Earth's hot interior, about 100 miles below the surface, and proceeds another 100 miles outwards through the ocean and air to the even hotter thermosphere at the edge of space. It includes the biosphere and is a dynamic physiological system that has kept our planet fit for life for over three billion years. I call Gaia a physiological system because it appears to have the unconscious goal of regulating the climate and the chemistry at a comfortable state for life. Its goals are not set points but adjustable for whatever is the current environment and adaptable to whatever forms of life it carries.

We have to think of Gaia as the whole system of animate and inanimate parts. The burgeoning growth of living things enabled by sunlight empowers Gaia, but this wild chaotic power is bridled by constraints which shape the goal-seeking entity that regulates itself on life's behalf. I see the recognition of these constraints to growth as essential to the intuitive understanding of Gaia. Important to this understanding is that constraints affect not only the organisms or the biosphere but also the physical and chemical environment. It is obvious that it can be too hot or too cold for mainstream life, but not so obvious is the fact that the ocean becomes a desert when its surface temperature approaches 15°C; when this happens, a stable surface layer of warm water forms that stays unmixed with the cooler, nutrient-rich waters below. This purely physical property of ocean water denies nutrients to the life in the warm layer, and soon the upper sunlit ocean water becomes a desert. This may be one of the reasons why Gaia's goal appears to be to keep the Earth cool.

You will notice I am continuing to use the metaphor of 'the living Earth' for Gaia; but do not assume that I am thinking of the Earth as alive in a sentient way, or even alive like an animal or a bacterium. I think it is time we enlarged the somewhat dogmatic and limited definition of life as something that reproduces and corrects the errors of reproduction by natural selection among the progeny.

I have found it useful to imagine the Earth as like an animal, perhaps because my first experience of serious science as a graduate was in physiology. It has never been more than metaphor – an *aide pensée*, no more serious than the thoughts of a sailor who refers to his ship as 'she'. Until recently no specific animal came into my mind, but always something large, like an elephant or a whale. Recently, on becoming aware of global heating, I have thought of the Earth more as a camel. Camels, unlike most animals, regulate their body temperatures at two different but stable states. During daytime in the desert, when it is unbearably hot, camels regulate close to 40°C, a close enough match to the air temperature to avoid having to cool by sweating precious water. At night the desert is cold, and even cold enough for frost; the camel would seriously lose heat if it tried to stay at 40°C, so it moves its regulation to a more suitable 34°C, which is warm enough. Gaia, like the camel, has several stable states so that it can accommodate to the changing internal and external environment. Most of the time things stay steady; as they were over the few thousand years before about 1900. When the forcing is too strong, either to the hot or the cold, Gaia, as a camel would, moves to a new stable state that is easier to maintain. She is about to move now.

Metaphor is important because to deal with,

understand, and even ameliorate the fix we are now in over global change requires us to know the true nature of the Earth and imagine it as the largest living thing in the solar system, not something inanimate like that disreputable contraption 'spaceship Earth'. Until this change of heart and mind happens we will not instinctively sense that we live on a live planet that can respond to the changes we make, either by cancelling the changes or by cancelling us. Unless we see the Earth as a planet that behaves as if it were alive, at least to the extent of regulating its climate and chemistry, we will lack the will to change our way of life and to understand that we have made it our greatest enemy. It is true that many scientists, especially climatologists, now see that our planet has the capacity to regulate its climate and chemistry, but this is still a long way from being the conventional wisdom. It is not easy to grasp the concept of Gaia, a planet able to keep itself fit for life for a third of the time the universe has existed, and until the IPCC (Intergovernmental Panel on Climate Change) sounded the alarm there was little inclination. I will try to provide an explanation that would satisfy a practical person like a physician. A complete explanation that would satisfy a scientist may be inaccessible, but the lack of it is no excuse for inaction.

I find explaining Gaia is like teaching someone how to swim or to ride a bicycle: there is much that cannot be put into words. To make it easier I will start at the shallow end with a simple question that illustrates the mind-wrenching difference between two equally important ways of thinking about the world. The first is systems science, which is about anything alive, whether an organism or an engineering mechanism while it is working; the second is reductionist science, the cause-and-effect thinking that has dominated the last two centuries of science. The question is: what has peeing to do with the selfish gene?

When I was a young man I was amazed by the number of euphemisms that existed for the simple but essential practice of passing urine. Doctors and nurses would ask you to 'produce a specimen' or 'pass some water' and often hand out a small container to make their request clear. In everyday speech we 'pumped the ship', 'sprung a leak' or 'shed the load' and we did it in 'the little boys' room' or the 'bathroom'. Sometimes we just 'spent a penny'.

Perhaps it was all a hangover from the nineteenth-century confusion over sex. It was not only impossible in polite speech to mention the genitals; the taboo applied also to their alternative uses. But as the outstanding American biologist George Williams observed in 1996, what an odd evolutionary economy to use the same organ for pleasure, reproduction and waste disposal. It was not until quite recently that I began to wonder if there might not be something deeper lurking behind this minor mystery. Why do we pee? Not so silly a question as it might seem. The need to rid oneself of waste products like excess salt, urea, creatinine and numerous other scraps of metabolism is obvious but only part of the answer. Perhaps we pee for altruistic reasons. If we and other animals did not pass urine some of the vegetable life of the Earth might be starved of nitrogen.

Is it possible that in the evolution of Gaia, the great Earth system, animals have evolved to excrete nitrogen as urea or uric acid instead of gaseous nitrogen? For us the excretion of urea represents a significant waste of energy and of water. Why should we evolve something to our disadvantage unless it was for altruistic reasons? Urea is the waste product of the metabolism of the meat, the fish, the cheese and the beans we eat; all are rich in protein, the stuff of life. We digest what we eat and break it down to its component chemicals; we do not take beef-muscle protein and use it in our own muscles. We build or replace our muscles and other tissue by assembling the component parts, the amino acids of the proteins, into fresh protein according to the plan in our DNA. To use the protein from beef directly to make our muscles would be like taking the parts of a tractor to repair a washing machine. The waste left over from this busy construction and deconstruction ultimately becomes urea, and we seem to have no option but to get rid of it as a dilute solution in water, urine.

Urea is a simple chemical, a combination of ammonia and carbon dioxide, or as an organic chemist would say, the di-amide of carbonic acid, NH_2CONH_2. Why did we and other mammals evolve to excrete our nitrogen in this form? Why not break down the urea into carbon dioxide, water and nitrogen gas? Much easier to excrete nitrogen by breathing it out, and it would save the water needed for excreting urea; oxidizing the urea would even add a little water, to say nothing of providing more energy.

Let us look at the figures: 100 grams of urea is metabolically worth 90 kilocalories or, if you prefer, 379 kilojoules. But if instead of being consumed it is passed in urine, more than 4 litres of water are needed to excrete the 100 grams of urea at a non-toxic dilution. Normally we excrete about 40 grams of urea daily in about 1.5 litres of water. Not much of a problem, you might think, but just consider animals living in a desert region short of food and water. If a mutant appeared that was able to metabolize urea to nitrogen, carbon dioxide and water, it would be at a considerable advantage and probably be able to leave more progeny than its urea-excreting competitors. According to a simplistic inter-pretation of Darwinian theory, selection would favour this mutant trait and it would spread rapidly and become the norm.

At this point a sceptical biochemist will say, 'Don't you realize that the products of ammonia or urea oxidation are all poisonous, and that is why we excrete nitrogen as urea?' My reply would be, 'Tell that to the bacteria that change nitrogen compounds into nitrogen gas and which are abundant in the soil and ocean.' More than this, a symbiosis with denitrifying organisms might be as good as or better than trying to metabolize urea ourselves.

So you see, urea is waste for us and wasting it loses valuable water and energy. But if we and other animals did not pee and breathed out nitrogen instead, there might be fewer plants and later we would be hungry. How on Earth did we evolve to be so altruistic and have such enlightened self-interest? Perhaps there is wisdom in the workings of Gaia and the way she interprets the selfish gene.

When I started working on Gaia fifty-five years ago, science was not as now a highly organized and often corporate enterprise. There was almost no forward planning or status reports, and there were almost never meetings to plan what to do next. There was no health and safety bureaucracy – we were expected to be, as qualified scientists, responsible for our own and our colleagues' safety. Most differently, science was done hands-on in the laboratory, not simulated on a computer screen in an office or a cubicle. In this idyllic environment it was possible to do an experiment to confirm or deny an idea. Sometimes the answer was a simple right or wrong, but on other occasions something equivocal. These 'don't knows' were what was led by serendipity to the revelation of something wholly unexpected, a real discovery.

So it might be with the idea of urea excretion. Thinking about nitrogen this way led me to wonder about the vexing problem of oxygen in the Carboniferous period some 300 million years ago. An important part of the evidence for Gaia comes from the abundance of atmospheric gases, such as oxygen and carbon dioxide; these are regulated at a level comfortable for whatever happens to be the current form of life. There are good experimental as well as theoretical grounds for thinking that the present percentage of oxygen in the atmosphere is about right. More than 21 per cent carries an increasing fire risk; at 25 per cent the probability of a blaze from a spark increases about tenfold. Andrew Watson and Tim Lenton have modelled the regulation of oxygen and have found the fire risk of dry vegetation to play an important part in the mechanism of oxygen regulation. Below 13 per cent there are no fires, and above 25 per cent they are so fierce that it seems impossible that forests could reach maturity. Imagine our surprise when the eminent geochemist Robert Berner proposed that in the Carboniferous period, about 300 million years ago, oxygen was 35 per cent of the atmosphere. His conclusion came from a model based on a thorough analysis of the composition of carboniferous rocks. He argued that at that time so much carbon was being buried, much of which we now see as the coal measures, that there had to be much more oxygen in the air to balance this greater rate of carbon burial.

My first reaction was that Berner must be wrong; I knew from the careful experiments made by my colleague Andrew Watson in the 1970s that fires in 35 per cent oxygen are almost as fierce as in pure oxygen. I was not impressed by laboratory experiments that suggested that twigs from trees did not readily inflame in 35 per cent oxygen; there is a world of difference between a laboratory simulation and a real forest fire, where its intense radiation dries out the wood in the path of the fire and where the winds drawn by the fire bring in fresh oxygen-rich air. Nor was I impressed by arguments that the huge dragonflies that existed at that time could not have flown without 35 per cent oxygen in the air. It is now realized that insects are unusually vulnerable to oxygen poisoning and that the Cretaceous dragonflies would have had no difficulty flying at our present oxygen levels. The argument went on until a friend, Andrew Thomas, an acoustic scientist and also a diver, suggested that maybe we were both right. Berner was right to claim that there was more oxygen and I was right to say it could not have been present at much over 25 per cent. All that was required was more nitrogen in the air. It is not the amount of oxygen that determines flammability, but its proportion in the mixture with nitrogen.

About 40 per cent of the nitrogen on Earth is now buried in the crust; perhaps in the

Cretaceous, that nitrogen had not yet been buried and existed in the air and so kept the proportion of oxygen safer for trees. We might also speculate that the microbial life of the Precambrian that preceded the appearance of trees and animals did not conserve nitrogen, so that it would have been present mainly as gas in the air.

These thoughts about nitrogen are wholly speculative, but I include them to illustrate the way that Gaia Theory has developed from ideas that were at first vague or from fruitful errors that were the seeds from which a truer account has emerged.

Self-regulation by feedback systems is notoriously difficult to explain in speech or writing. In his recent book, *Facing Gaia* and in his papers with my colleague, Tim Lenton, Bruno Latour comes closer to a verbal explanation of Gaia than any other writer I have yet encountered.

For myself, I resort to metaphor and continue to say that the Earth system is alive.

So, let us go further now and try to sense Gaia by looking at the Earth from outside as a whole planet. Imagine a spacecraft manned by intelligent aliens who are looking at the solar system from space. They would have aboard their ship instruments powerful enough to show the travellers the chemical composition of every planet's atmosphere. From this analysis and nothing more, their automated instruments would tell them that the only planet with abundant life was the Earth; more than that, they would say that the life form was carbon-based and was sufficiently advanced to have an industrial civilization. There is nothing science fictional about the instrument itself; a small telescope with an infra-red spectrometer and a computer to control them and analyse their observations would do. They would see methane and oxygen coexisting in the upper air of the Earth, and the ship's scientist would know that these gases were reacting in the bright sunlight and that therefore something on the ground must be making large quantities of them both. The odds against this happening by chance inorganic chemistry are near infinity. They would conclude that our planet is a rich habitat for life, and the presence of CFCs would suggest a civilization unwise enough to have allowed their escape.

In the 1960s I was a contractor designing instruments for NASA's planetary exploration team, and thoughts like these led me to propose planetary atmospheric analysis for the detection of life on Mars. I argued that if there was life on Mars it would have to use the atmosphere as a source of raw materials and as somewhere to deposit its wastes; this would change the atmospheric composition and make it recognizably different from that of a dead planet. I saw the Earth, rich with life, as the contrasting planet, and I used the eminent scientist G. E. Hutchinson's authoritative review of biogeochemistry as my source of information on the sources and sinks for the gases of the air. He reported methane and nitrous oxide as biological products, and nitrogen, oxygen and carbon dioxide as massively changed in abundance by organisms. At the time, none of us knew much about the composition of Mars's atmosphere, but in 1965 Earth-based infrared astronomy revealed the Mars atmosphere to be composed almost entirely of carbon dioxide and close to chemical equilibrium; according to my proposal it was therefore probably lifeless – not a popular conclusion to give my sponsors. Turning aside from life detection, I wondered what could be keeping our chemically unstable atmosphere in a dynamic steady state and the Earth always apparently habitable. Moreover, the continuity of life requires a tolerable climate despite a 37 per cent increase of solar luminosity since the Earth formed. Together, these thoughts led me to the hypothesis that living organisms regulate the climate and the chemistry of the atmosphere in their own interest, and in 1969 the novelist William Golding proposed Gaia as its name. A few years later, I started collaborating with the eminent American biologist Lynn Margulis, and in our first joint paper we stated: the Gaia Hypothesis views the biosphere as an active, adaptive control system able to maintain the Earth in homeostasis.

From its beginning in the 1960s, the idea of the global self-regulation of climate and chemistry was unpopular with both Earth scientists and life scientists. At best, they found it unnecessary as an explanation of the facts of life and the Earth; at worst, they condemned it outright in scathing terms. The only scientists who welcomed the idea were a few meteorologists and climatologists. Some biologists soon challenged the hypothesis, arguing that a self-regulating biosphere could never have evolved, since the organism was the unit of selection, not the biosphere. I was fortunate to have that fine and clear author Richard Dawkins as the advocate for the Darwinian opposition to Gaia; it was painful but in time I found myself agreeing with him that Darwinian evolution, as it was then understood, was incompatible with the Gaia Hypothesis. I did not then doubt Darwin, so what was wrong with the Gaia hypothesis? I knew that the constancy of climate and of the chemical composition of the air were good evidence for a self-regulating planet. Moreover, the concept of Gaia is fruitful, and it led me to discover the natural molecular carriers of the elements sulphur and iodine: dimethyl sulphide (DMS) and methyl iodide. Several years later in 1986, while collaborating with colleagues in Seattle, we made the awesome discovery that DMS from ocean algae was connected with the formation of clouds and with climate. We were moved to catch a glimpse of one of Gaia's climate-regulation mechanisms, and we were indebted to the climate-science community who took us seriously enough to award to the four of us, Robert Charlson, M. O. Andreae, Steven Warren and me, their Norbert Gerbier Prize in 1988.

To return to the arguments with the Darwinists, it occurred to me in 1981 that Gaia was the whole system – organisms and material environment coupled together – and it was this huge Earth system that evolved self-regulation, not life or the biosphere alone. To test this idea I composed a computer model of dark- and light-coloured plants competing for growth on a planet in progressively increasing sunlight. It was no more than a simulation of the world, but the running program showed the imaginary world regulating its temperature close to the optimum for daisy growth and over a wide range of heat outputs from its star. This model, which I called Daisyworld, was unusual for an evolutionary model made from coupled differential equations; it was stable, insensitive to initial conditions and resistant to perturbation.

Daisyworld models a planet like the Earth, orbiting a star like our sun. On Daisyworld there are only the two plant species, and they both compete for living space as any plants would do. When the sun is younger and cooler, so is the model planet, and at that time the dark daisies flourish. Only at the hottest places near the equator are light daisies found. This is because dark daisies absorb sunlight and keep themselves, their region and the whole planet warm. As the star heats up, the dark daisies living in the tropics are displaced by light daisies, because the light ones reflect sunlight and so are cooler; they also cool their region and the whole planet. As the star continues to warm, the light daisies displace the dark, and through their competition for space the planet always stays near to the ideal temperature for life. Eventually, the star grows so hot that even light daisies can no longer survive and the planet becomes a lifeless ball of rock.

The model is no more than a caricature, but think of it like that splendid map of the London Tube system – not good as a guide to the streets of London, but ideal for finding your way around the Tube system of that bustling city. Daisyworld was invented to show that Darwin's theory of evolution from natural selection is not contrary to Gaia theory, but part of it.

The main reaction of biologists and geologists to Daisyworld was, as good scientists, to try to falsify it, and this they did repeatedly, with increasing irritation, but none succeeded. To answer some of these critics I made models much richer in species than Daisyworld. They included many different types of plant, rabbits to graze them and foxes as predators. They were just as stable and self-regulating as Daisyworld. My friend Stephan Harding has made models of whole ecosystems complete with food webs and used them to enlighten our understanding of biodiversity. The persistence of the critics made me realize that Gaia would not be taken as serious science until eminent scientists approved of it in public. In 1995 I started dialogues with John Maynard Smith and William Hamilton, both of whom were prepared to discuss Gaia as a scientific topic but neither of whom could see how planetary self-regulation could evolve through natural selection. Even so, Maynard Smith gave unstinting support to my friend and colleague Tim Lenton when the latter wrote a seminal article in *Nature* called 'Gaia and Natural Selection'. In it he described the several ways that the Earth keeps to its goal of sustaining habitability for whatever life forms happen to be its inhabitants. Hamilton wondered in a joint paper with Lenton, with the provocative title 'Spora and Gaia', if the need for organisms to disperse was the link that connected ocean algae with climate. In 1999 Hamilton said in a television programme, 'Just as the observations of Copernicus needed a Newton to explain them, we need another Newton to explain how Darwinian evolution leads to a habitable planet.'

Then, at least in Europe, the ice began to melt, and at a meeting in Amsterdam in 2001 – at which four principal global-change organizations were represented – more than a thousand delegates signed a declaration that had as its first main statement: 'The Earth System behaves as a single, self-regulating system comprised of physical, chemical, biological and human components.'

These words marked an abrupt transition from a previously solid conventional wisdom in which biologists held that organisms adapt to, but do not change, their environments and in which Earth scientists held that geological forces alone could explain the evolution of the atmosphere, crust and oceans. We should recall at this point the trials of that eminent biologist Eugene Odum, who in the 1960s saw an eco-system as an entity like Gaia. So far as I am aware, none of the biologists who stridently rejected Odum's concept have admitted that they were wrong.

The Amsterdam Declaration was an import-ant step towards the adoption of Gaia Theory as a working model for the Earth; however, terri-torial divisions and lingering doubts kept the declaring scientists from stating the *goal* of the self-regulating Earth, which is, according to my theory, to sustain habitability. This omission allows scientists to pay lip service to Earth Sys-tem Science (ESS), or Gaia, but continue to model and research in isolation as before. This natural and human tendency of scientists to resist change would not ordinarily have mat-tered: eventually the strings of habit would have broken and geochemists would have started to think of the biota as an evolving and responding part of the Earth, not as if life were merely a passive reservoir like the sediments or the oceans. Eventually also biologists would have thought of the environment as something that organisms actively changed and not as something fixed to which they adapted. But unfortunately, while scientists are slowly changing their minds, we of the industrial world have been busy changing the surface and atmosphere. Now humanity and the Earth face a deadly peril, with little time left to escape. If the middle management of science had been somewhat less reactionary about Gaia, we might have had twenty more years in which to resolve the much more difficult human and political decisions about our future.

The Life History of Gaia

Life on Earth began between three and four billion years ago; we can only guess the date, since there are so few unequivocally dated fossils to be found. At this early time the sun was probably 23 per cent less luminous than it is now. We think that the Earth was mainly covered by ocean and there were only small continents. It would have been kept warm enough for water to stay liquid and for life to start through the presence of abundant carbon dioxide in the atmosphere, perhaps thirty times more than now, and it may have been a darker planet than now, because there was less land and possibly fewer clouds. Once photosynthesis evolved it would have used the carbon dioxide as its carbon source and by so doing decreased its abundance in the air.

We could look at this as a reverse greenhouse effect that presented early life with problems like the greenhouse warming that we face today, but for early life the threat was cooling or freezing, not warming. We think early life resolved this problem through the evolution of organisms called methanogens, which are still around in our guts and anywhere there is a lack of oxygen. These 'detritophores' live by decomposing the bodies of deceased photosynthesizers and other organisms; the main products of their decomposition are the gases methane and carbon dioxide. Methane is twenty-four times as potent a greenhouse gas as carbon dioxide, and when its atmospheric abundance was about 100 ppm in the early Earth's atmosphere it would easily have kept our infant planet warm enough for life. This idea, first mentioned in my book *The Ages of Gaia* in 1988, is slowly becoming the conventional wisdom among geochemists.

Once Gaia came into existence as a planetary system (and I think that this would have been some time after life itself had started) it would have changed the atmosphere from one dominated by carbon dioxide to one dominated by methane. If a Gaian feedback system had evolved, this ancient world of bacteria could have been dynamically stable and resilient against perturbation, but it still would have been vulnerable to catastrophes, such as planetesimal impacts or huge volcanic outbursts. After Gaia had emerged, an event of this kind would have sterilised the Earth, because methane would have vanished rapidly from the air and the Earth could have frozen. But in those earlier times recovery was automatic, as carbon dioxide vented into the air from volcanoes and so made a new gaseous greenhouse that rewarmed the planet. There might have been enough survivors to rebuild the smelly septic tank world of infant Gaia.

Things are quite different now. Any catastrophe that caused Gaia's regulation system to fail would lead to a hot and arid Earth with no natural means of returning to its cooler state.

Simple models of Gaia are stable and not easily perturbed, but only if more than a critical mass of life is present on the model planet. The model usually comes to equilibrium with 70 to 80 per cent of the planetary surface inhabited, the remainder assumed to be barren or sparsely populated desert or ocean. If a plague or some other mishap destroys more than 70 to 90 per cent of the population, the temperature and chemical composition cease to be regulated and the model system swiftly drops to the equilibrium state of the dead planet.

The vulnerability of these model systems to upsets depends upon the intensity of the stress the planet is undergoing before the disturbance occurs. With a model of the Earth two billion years ago I found that almost all of the living organisms could be eliminated without disturbing the planetary climate. At this time the Earth was briefly passing through its 'Goldilocks' stage, when the heat from the sun was just right for life and little or no temperature regulation was needed. This may have been why one of the great crises of Gaia's existence, the appearance of oxygen as a dominant atmospheric gas, passed without deadly consequences. It happened when the climate of the solar system was benign. At the beginning, over three billion years ago, the sun was too cool for comfort – now it is too hot.

The appearance of oxygen was an event as important in Gaian history as puberty is in humans. It drove the evolution of more complex living cells, the eukaryotes and eventually the huge assemblies of living cells that make up plants and animals. Not least, it allowed the Earth to retain its oceans by acting as a barrier against the escape of hydrogen to space. For over a billion years after oxygen appeared, the evolution of life on Earth passed through something like a dark age, with little or no historical evidence. This period, the proterozoic, was one where life was still unicellular, and it left behind in the geological record almost nothing in fossil form.

Our view of the Earth's past is like that of a landscape from a mountain viewpoint. Apart from a few other snowy distant peaks, large forests and lakes, nothing detailed is discernible beyond a mile or so; the history of the British Isles in the ice ages of the Pleistocene falls in this discernible range. During the brief warm interglacials, it seems to have been an unbroken, shore-to-shore carpet of trees, a broad-leafed temperate forest ecosystem, small compared with the huge tropical rainforests of today, but like them diverse in its range of species. The carpet of trees covered nearly all of the land, including the mountain areas that are now treeless; indeed, what is often spoken of now as wilderness was then covered in trees. Grazing animals would have made a few clearings and forest paths, but these would represent only a tiny fraction of the whole. A bird flying high over the British Isles would have seen a densely packed forest extending to the horizon, just like a present-day aerial photograph of Amazonia.

I find it remarkable that such a verdant scene has alternated more than twenty times, with much longer periods of tundra and glaciers that, seen from above, would have looked like Greenland today. The long ice ages swept away the trees and all but sterilized the land; yet when the climate warmed for the short interglacials, life returned anew and in much the same way every time. The frostbitten extremities of the Earth healed well when a warmer climate came.

As a geophysiologist, I look on these cold and warm events as a series of experiments. Trees and other plants were seeded onto the warm but sterile land that was set free as the glaciers retreated, and they rapidly grew until there was confluent forest cover. Then the experimental region was put in the deep freeze of a glaciation until it was time for a repeat. It was a good series of experiments, and in the many repetitions the results varied only by a small amount. A botanist, for example, would notice variations in the organisms present: sometimes there would be mainly oak, while in other, colder periods, alders, birch and conifers would predominate.

I suspect, but do not know, that the biodiversity – that is, the number of different species present in a defined area – would also have changed. Stable unchanging climates lasting for several thousand years tend to reduce diversity, but when the climate changes to either hotter or colder by a small amount the first response is an increase in biodiversity. This is because the new conditions give rare species a chance to flourish while the established ones have not had time to decline. When the climate stabilizes again, survivors of the past regime may die out and biodiversity diminish again. Of course, biodiversity falls almost to zero in the impoverished environment of an ice cap, but it is important to keep in mind that biodiversity and environmental quality are not simply proportional.

A planetary physician would look on biodiversity as a symptom, a response to change. He would recognize that what is a rare species in one state becomes a common one in another. So rich biodiversity is not necessarily something highly desirable and to be preserved at all costs. A red, flushed and sweaty skin is our physiological response to overheating, and the biodiversity of a tropical forest like Amazonia may be the Earth's response to the heat of the present interglacial. Neither of these states is worth preserving as a long-term goal, and evolution would change them into something more stable. I suspect that the capacity to become biodiverse has evolved because, in the real world of Gaia, change is always happening and is usually driven from outside by small alterations in the seating arrangements of the solar system and in the output from the sun. When there is a climate change, dormant seeds, rare plants, or seeds drifting in on the wind, or on the feet of birds, have a better or worse chance to grow; if better, they flourish and compete with the native species until they become a stable part of the ecosystem. During the period of competition biodiversity is increased, but it declines again as the ecosystem adapts to the new conditions.

We have become so concerned over the fate of the rare tree, especially if it produces a drug that might cure cancer, and about rare and beautiful animals and birds; we have become so excited by these collectables that we have lost sight of the forest itself. But Gaia's automatic response to adverse change is driven by the changes in the whole forest ecosystem, not by the presence or absence of rare species alone. Niches vacated by extinction do not stay empty, and like the great rentier that she is, they are rapidly occupied; her rent, the cash flow of elements, is just as well paid by dull and abundant plants as it is by rarities – as with the human ecosystem of London, which displays its exotica in the habitats of Hampstead, Notting Hill, and Islington.

But what of the glaciations, when it grows really cold and ice begins to scrape away the soil and destroy almost all life? Why does Gaia not resist this adverse change? The answer lies, I think, in a long-term, whole-planet view. As the aeons have passed, the sun has remorselessly grown hotter; that is the nature of the nuclear furnaces that power stars, and as they age they increase their heat output and eventually die in a burst of fire. In order to sustain an equable climate the Earth system has evolved several air-conditioning mechanisms. Vegetation growing on the land and floating in the sea uses carbon dioxide that it removes from the air, and this lessens the carbon-dioxide abundance and its greenhouse effect; another mechanism is the production by marine organisms of gases that, when oxidized in the air, make the tiny particles called cloud condensation nuclei, without which water in the air would not condense as the droplets that clouds are made of. Without clouds, the Earth would be much hotter.

The period we are now in is close to a crisis point for Gaia. The sun is now too hot for comfort, but most of the time the system has managed to pump down carbon dioxide sufficiently and to produce enough white reflecting ice and clouds to keep the Earth cool and to maximize the occupancy of the Earth's niches. But to do so the regions above 45° north and below 45° south of the equator have had to be sacrificed. This is not as large a loss to Gaia as it is to humans. These polar regions occupy less than 30 per cent of the Earth's surface, and their white reflecting surfaces powerfully assist cooling.

During an ice age, so much water is locked up in the glaciers of the polar regions that the sea level drops by 120 metres. Consequently, a vast area of land emerges from the sea, and much of it is in the tropics; Tim Lenton reminded me that the land released by the fall of sea level was equal in area to that covered by ice. The loss of productivity in the temperate and polar latitudes is more than compensated for by the increase in land life in the tropics and in the cooler oceans. Although there is a smaller area of ocean in an ice age, it is more productive, because cold water favours the growth of the primary producers, the photosynthetic algae. As I mentioned earlier, a warm ocean is, perversely, nowhere near as productive as a chilly one. The colder waters are the dense forests of the sea, rich in life and helping to keep the Earth cool by producing clouds and by pumping down carbon dioxide.

The Senescence and Death of Gaia

The energy source of the solar system is the sun. This nuclear furnace has now operated for four and a half billion years and will continue for about another five billion, when its supply of fuel – hydrogen and helium – runs out. In the long term the sun is not renewable, but in our terms it can be taken as so. The sun is a remarkably steady and reliable source of light and heat, and the supply is 1.35 kilowatts of energy for every square metre of the Earth that is in direct unimpeded sunlight.

Because the sun grows hotter, the heat received by the Earth now is more than it was when life began over three billion years ago. Yet most textbooks and television programmes on science will tell you that the Earth, like

Goldilocks, is a planet that happened to be born at exactly the right distance from the sun, and this is why conditions on Earth are exactly right for life. This pre-Gaia statement is wrong, and only for a brief period in the Earth's history was the sun's warmth ideal for life, and that was about two billion years ago. Before this it was too cold for comfort and afterwards it has progressively grown too hot. In the very long term, solar warming is a far greater problem for life than our present-day battle with man-made global heating.

In about one billion years, and long before the sun's life ends, the heat received by the Earth will be more than two kilowatts per square metre, which is more than the Gaia we know can stand; she will die from overheating. Gaia regulates its temperature at what is near optimal for whatever life happens to be inhabiting it. But, like many regulating systems with a goal, it tends to overshoot and stray to the opposite side of its forcing. If the sun's heat is too little the Earth tends to be warmer than ideal; if too much heat comes from the sun, as now, it regulates on the cold side of ideal. This is why the usual state of the Earth at present is an ice age. The recent crop of glaciations the geologists call the Pleistocene is, I think, a last desperate effort by the Earth system to meet the needs of its present life forms. The sun is already too hot for comfort. The low level of carbon dioxide gives a measure of the problems faced by Gaia during an ice age; planetary life pumps down carbon dioxide from the air until it reaches levels as low as 180 ppm. This is half of what is in the air now and is too little for some plants to grow well. Michael Whitfield and I calculated, in 1981, that in less than 100 million years the sun's heat will be too much for the Earth to regulate at its current state, and it will be forced to move to a new hot state inhabited by a different biosphere. The brief interglacials, like now, are, I think, examples of temporary failures of ice-age regulation. These ideas were taken up and extended by Jim Kasting and Ken Caldeira in 1992 and by Tim Lenton and Werner von Bloh in 2001.

Looked at on this long-term and large scale we sense that our adding carbon dioxide to the air and soon doubling its abundance is seriously destabilizing an Earth system already struggling to maintain the desired temperature. By adding greenhouse gases to the air and by replacing natural ecosystems, like forests, with farmland we are hitting the Earth with a 'double whammy'. We are interfering with temperature regulation by turning up the heat and then simultaneously removing the natural systems that help to regulate it. What we are now doing is uncannily like the series of foolish actions that led to the Chernobyl nuclear reactor accident. There the engineers turned up the heat after they had disabled the safety systems, and it should have been no surprise that the reactor ran into rapid overheating and caught fire.

Climatologists now think that we are perilously close to the threshold beyond which adverse change sets in; change that is, on a human timescale, irreversible. The Earth does not catch fire, but it becomes hot enough to melt most of the Greenland ice and some of the West Antarctica ice; enough water will then be added to the world oceans to raise sea levels by fourteen metres. It is sobering to think that nearly all of the present great centres of population are currently below what could be the ocean surface in a mere blink of geological time.

It would be wrong to leave this account of Gaia without touching again on the fact that she is old and has not very long to live. As the sun grows ever hotter it will, in Gaia's terms, soon become too hot for animals and plants and many of the microbial forms of life. I think it unlikely that heat-tolerant bacteria, thermophiles living in the oases of a desert world, would be abundant enough to form the critical mass of living things needed for Gaia. It is also unlikely that the kind of Earth we know now would last even a fraction of those billion years. The harm done by a planetesimal impact, or even by a future industrial civilization, may drive Gaia first to one of the hotter and temporarily stable states, and finally to total failure.

Growing old is not as bad as is sometimes imagined. When I was in my teenage years it seemed then that by now I would be feeble, depressed and barely even half-witted. Some, but not all, of these premonitions have come true, and although I can walk and climb a modest slope, walking at that speed over mountains is no longer an option. But somehow I learnt that life begins anew at each decade; it certainly, for me, began afresh at each decade from the age of twenty onwards. As with a butterfly, the long years as a grub and then a pupa are over, and as the poet Edna St Vincent Millay said:

> My candle burns at both ends;
> It will not last the night;
> But, ah, my foes, and oh, my friends –
> It gives a lovely light.

So it is with Gaia. The first aeons of her life were bacterial, and only in her equivalent of late middle age did the first meta-fauna and meta-zoa appear. Not until her eighties did the first intelligent animal appear on the planet. Whatever our faults, we surely have enlightened Gaia's seniority by letting her see herself from space as a whole planet while she was still beautiful. Unfortunately, we are a species with schizoid tendencies, and like an old lady who has to share her house with a growing and destructive group of teenagers, Gaia grows angry, and if they do not mend their ways she will evict them.

A Personal View of Environmentalism

The concept of Gaia, a living planet, is for me the essential basis of a coherent and practical environmentalism; it counters the persistent belief that the Earth is a property, an estate, there to be exploited for the benefit of humankind. This false belief that we own the Earth, or are its stewards, allows us to pay lip service to environmental policies and programmes but to continue with business as usual. A glance at any financial newspaper confirms that our aim is still growth and development. We cheer at any new discovery of gas or oil deposits and regard the current rise in petroleum prices as a potential disaster, not a welcome curb on pollution. Few, even among climate scientists and ecologists, seem yet to realize fully the potential severity, or the imminence, of catastrophic global disaster; understanding is still in the conscious mind alone and not yet the visceral reaction of fear. We lack an intuitive sense, an instinct, that tells us when Gaia is in danger.

So how do we acquire, or reacquire, an instinct that recognizes not only the presence of the great Earth system but also its state of health? We do not have much to go on because the concepts of intuition and instinct tended to be ignored, or at best regarded as flaky and dubious, during the last two centuries of triumphant reductionism. In the twenty-first century we are somewhat freer to wonder about ideas like instinct and intuition, and it seems probable that long ago in our evolutionary history, when our ancestors were simple aquatic animals, we had already evolved an ability instantly to distinguish anything alive within the mainly inorganic ocean. This primeval instinct would have been supremely important for survival, since living things can be either edible, lovable or lethal. It is likely to be part of our genetic coding and hard wired into our brains so that we still have it in full strength. We do not need a doctorate in biology to distinguish a beetle from a stone, or a plum from a pebble. But, because of the circumscribed nature of its origins, the instinctive recognition of life is limited by the range of our senses and does not work for things smaller or larger than we can see. We recognize a paramecium as alive, but only when we can see it through a microscope. Even biologists, when they think of the biosphere, too often ignore all things smaller than can be seen with the naked eye. My friend and collaborator Lynn Margulis more than anyone has stressed the primary importance of microorganisms in Gaia, and she summarizes her thoughts in the book she wrote in 1986 with Dorian Sagan, *Microcosmos*. The Earth was never seen as a whole until astronauts viewed it for us from outside, and then we saw something very different from our expectation of a mere planet-sized ball of rock existing within a thin layer of air and water. Some astronauts, especially those who travelled as far as the moon, were deeply moved and saw the Earth itself as their home. Somehow we have to think like them and expand our instinctive recognition of life to include the Earth.

★

The ability instantly to recognize life, and other instincts, like the fear of heights and snakes, are part of our long evolutionary history, but there is another kind of instinct that is not innate but grows from childhood conditioning. The Jesuits discovered that a child's mind could be moulded to accept their faith, and that once done the child retained faith as an instinct throughout life; similar but different moulds fix lifelong tribal and national loyalty. The mind of a child is even plastic enough to be shaped to follow faithfully something as trivial as a football team or as potentially sinister as a political ideology. Abundant experience of this kind suggests that we could, if we chose, make Gaia an instinctive belief by exposing our children to the natural world, telling them how and why it is Gaia in action, and showing that they belong to it.

The founders of the great religions of Judaism, Christianity, Islam, Hinduism and Buddhism lived at times when we were far less numerous and lived in a way that was no burden to the Earth. Those holy men would have had no inkling of the troubled state of the planet a thousand or more years later, and their concern, rightly, would have been for human affairs. Rules and guidance were needed for individual, family and tribal good behaviour; we were the human family growing up in the natural world of Gaia and, like children, we took our home for granted and never questioned its existence. The success of these religious backgrounds is measured by their persistence as faiths and guides over more than a thousand years of further population expansion. When I was a child I was marinated in Christian belief, and still it unconsciously guides my thinking and behaviour. Now we face the consequences of fouling our planetary home, and new hazards loom that are much more difficult to understand or cope with than the tribal and personal conflicts of the past. Our religions have not yet given us the rules and guidance for our relationship with Gaia. The humanist concept of sustainable development and the Christian concept of stewardship are flawed by unconscious hubris. We have neither the knowledge nor the capacity to achieve them. We are no more qualified to be the stewards or developers of the Earth than are goats to be gardeners.

Perhaps Christians need a new Sermon on the Mount that sets out the human constraints needed for living decently with the Earth, and which spells out the rules for its achievement. I have long wished that the religions and the secular humanists might turn to the concept of Gaia and recognize that human rights and needs are not enough; those with faith could accept the Earth as part of God's creation and be troubled by its desecration.

Important concepts like God or Gaia are not comprehensible in the limited space of our conscious minds, but they do have meaning in that inner part of our minds that is the seat of intuition. Our deep unconscious thoughts are not rationally constructed; they emerge fully formed as our conscience and an instinctive ability to distinguish good from evil. Perhaps this is why the early Quakers knew that the still, small voice within does not come from conscious reckoning. Our conscious rational minds are no more capable of deep thought than is the tiny screen of a contemporary mobile telephone able to present in its full glory a Vermeer painting. The extraordinary power of our unconscious minds is expressed in what we see as mundane things like walking, riding a bicycle or catching a ball. We would fail utterly to do any of these things by conscious thought; their automatic and instinctive achievement requires long and often tedious training. The same is true of inventors who, after long apprenticeship to their craft, become inspired to imagine and then construct devices that reveal emergence when they are switched on; physicists in a similar way exploit the incredible mysteries of quantum phenomena despite having no conscious understanding.

The history of science shows that we need to keep what is good in past interpretation of the world and merge in new knowledge as it appears. Newton's understanding enlightened physics for three hundred years. Einstein's relativity did not cast out Newtonian physics, it extended it. In a similar way, Darwin's great vision of evolution has raised biology from a cataloguing activity into a science, but now we are beginning to see Darwinism is incomplete. Evolution is not just a property of organisms – what evolves is the whole Earth system with its living and non-living parts existing as a tight coupled entity. It is foolish to think that we can explain science as it evolves, rationally and consciously. We have to use the crude tool of metaphor to translate conscious ideas into unconscious understanding. Just as the metaphor, a living Earth, used to explain Gaia, was wrongly rejected by reductionist scientists, so it may be wrong of them also to reject the metaphors and fables of the sacred texts. Crude they may be, but they serve to ignite an intuitive understanding of God and creation that cannot be falsified by rational argument.

As a scientist I know that Gaia Theory is provisional and likely to be displaced by a larger and more complete view of the Earth. But for now I see it as the seed from which an instinctive environmentalism can grow; one that would instantly reveal planetary health or disease and help sustain a healthy world.

Green thoughts and ideas are as diverse and competitive as the plants of a forest ecosystem and, unlike the plants, they do not even share the spectral purity of the colour of chlorophyll. Green thoughts range from shades of red to shades of blue. The totalitarian greens, sometimes called ecofascists, would like to see most other humans eliminated in genocide and so leave a perfect Earth for them alone. At the other end of the spectrum are those who would like to see universal human welfare and rights, and somehow hope that luck, Gaia or sustainable development will allow this dream to come true. Greens could be defined as those who have sensed the deterioration of the natural world and would like to do something about it. They share a common environmentalism but differ greatly in the means for its achievement.

The root of our problems with the environment comes from a lack of constraint on the growth of population. There is no single right number of people that we can have as a goal: the number varies with our way of life on the planet and the state of its health. It has varied naturally from a few million when we were hunters and gatherers to a fraction of a billion as simple farmers; but now it has grown to over seven billion, which is wholly unsustainable in the present state of Gaia, even if we had the will and the ability to cut back.

If we could go back to, for example, 1840 and start again we might be able to reach a stable population of six billion if we were guided from the beginning by a proper understanding of the Earth. We would know that fossil-fuel combustion needed limiting and that cattle and sheep farming use far too much land and cannot be sustained, and that arable farming, with pigs and chickens as food animals consuming mainly vegetable waste, would be a better way to go. It might even be possible to sustain ten billion or more living in well-planned, dense cities and eating synthesized food.

If we can overcome the self-generated threat of deadly climate change, caused by our massive destruction of ecosystems and global pollution, our next task will be to ensure that our numbers are always commensurate with our and Gaia's capacity to nourish them. Personally I think we would be wise to aim at a stabilized population of about half to one billion, and then we would be free to live in many different ways without harming Gaia. At first this may seem a difficult, unpalatable, even hopeless task, but events during the last century suggest that it might be easier than we think. Thus in prosperous societies, when women are given a fair chance to develop their potential they choose voluntarily to be less fecund. It is only a small step towards a better way of living with Gaia, and it has brought with it problems of a distorted age structure in society and dysfunctional family life, but it is a seed of optimism from which other voluntary controls could grow and surely far better than the cold concept of eugenics that withered in its own amorality. In the end, as always, Gaia will do the culling and eliminate those that break her rules. We have the choice to accept this fate or plan our own destiny within Gaia. Whatever we choose to do we have always to ask, what are the consequences?

The regulation of fecundity is part of population control, but the regulation of the death rate is also important. Here, too, people in affluent societies are choosing voluntarily seemly ways to die. Traditionally, hospitals have for the elderly been places for dying in comparative comfort and painlessness; the hospice movement has served to set standards and make this otherwise unmentionable role of the health systems acceptable. According to Hodkinson, in his book *An Outline of Geriatrics,* about 25 per cent of the elderly entering hospitals die within two months. Now that the Earth is in imminent danger of a transition to a hot and inhospitable state, it seems amoral to strive ostentatiously to extend our personal lifespan beyond its normal biological limit of about one hundred years. When I was a young post-doctoral fellow at Harvard Medical School in Boston an eminent paediatrician complained of the huge, more than tenfold, disparity between funds given for cancer research and those given for childhood disease; I suspect that it still exists.

We have severed nearly all the natural physical constraints on the growth of our species: we can live anywhere from the Arctic to the tropics and, while they last, our water supplies are piped to us. We are only just beginning to realize that pandemics may evolve naturally as a negative feedback on population growth. If we are to continue as a civilization that successfully avoids natural catastrophes, we have to make our own constraints on growth and make them strong and make them now.

Over half the Earth's people live in cities, and they hardly ever see, feel or hear the natural world. Therefore our first duty should be to convince them that the real world is the living Earth and that they and their city lives are a part of it and wholly dependent on it for their existence. Our role is to teach and to set an example by our lives. In certain ways my long-time friend Edward Goldsmith was one of the few who tried to live and think as a deep ecologist. His erudite and thought-provoking book *The Way* is essential reading for anyone who wants to know more about green philosophy.

I would like to see us use our technical skills to cure the ills of the Earth as well as those of humans. This is not surprising since my first experience in science was 25 years in medical research. I cannot stand aside while civilization drinks itself to death in fossil fuels. And this is why I regard nuclear energy as a needed remedy.

The green community should have been reluctant to found lobbies and political parties; both are concerned with people and their problems. Our task as individuals is to think of Gaia first. In no way does this make us inhuman or uncaring; our survival as a species is wholly dependent on Gaia and on our acceptance of her discipline.

I am often asked, 'What is our place in Gaia?' To answer we need to look back a long time ago in human history to when we were an animal, a primate, living within Gaia and different from other species only in unimportant ways. Our role then was like theirs, to recycle carbon and other elements. We lived on an omnivorous diet and returned to the air as carbon dioxide the carbon collected in their lifetimes by our food animals and plants. We had our niche in the evolutionary system, and our numbers were probably not more than a million.

As intelligent predators, we were equipped with useful brains and hands and could alter the boundaries of our niche in ways that were unavailable to other animals. We could throw stones, use simple stone and wood tools, and do it better than other primates.

Many animals, even insects like bees and ants, can communicate. They use alarms and mating calls and pass on detailed information about the size, direction and distance of food sources. We humans were fortunate to acquire through a mutation the ability to modulate our voices sufficiently for a primitive spoken language. This change was as profound for us as primitive people as the invention of the computer or mobile phone has been for modern humans. The members of the tribe could share experiences; they could plan ahead against drought and famine and guard against predators. We were by then the emerging *Homo sapiens* and may have been the first animals consciously to modify the environment for their own benefit. Most remarkably, we used natural fires started by lightning for cooking, clearing land and hunting.

The innocent among the urban intelligentsia think and talk of early humans as living in harmony with the natural world. Some of them go further and gather funds to preserve what they see as natural communities living in remote regions, such as the tropical forests. They see the modern world as clever but bad and these simple lifestyles as natural and good. They are wrong. We should not think of early humans as better or worse than we are; indeed, they were probably very little different.

Others consider us superior because of our cultured ways and intellectual tendencies; our technology lets us drive cars, use word processors and travel great distances by air. Some of us live in air-conditioned houses and we are entertained by the media. We think that we are more intelligent than stone-agers, yet how many modern humans could live successfully in caves, or would know how to light wood fires for cooking, or make clothes and shoes from animal skins or bows and arrows good enough to keep their families fed? I am indebted to Jerry Glynn and Theodore Gray for making this point in their guidebook for users of the computer program *Mathematica*, a mathematics processor. Using as an example the fact that modern children can hardly add a column of numbers without a calculator, they observe that this is no bad thing, since each stage of human development brings with it a full measure of skills exchanged for others no longer needed; stone-agers were probably as fully occupied with living as we are.

One group of these early humans migrated to Australia at a time when the sea levels were much lower than now and the journey by boat or raft was probably neither long nor difficult. From this group are descended the modern Australian aboriginals, often claimed to be an example of natural humans at peace with the Earth. Yet their method of clearing forests by fire may have destroyed the natural forests of the Australian continent as surely as do modern men with chainsaws. Peace on you Aboriginals; you individually are no worse and no better than we are, it is just that we are power-assisted and more numerous.

Through Gaia I see science and technology as traits possessed by humans that have the potential for great good and great harm. Because we are part of, and not separate from Gaia, our intelligence is a new capacity and strength for her as well as a new danger. Evolution is iterative, mistakes are made, blunders committed; but in time that great eraser and corrector, natural selection, usually keeps a neat and tidy world. Perhaps our and Gaia's greatest error was the conscious abuse of fire. Cooking meat over a wood fire may have been acceptable, but the deliberate destruction of whole ecosystems by fire merely to drive out the animals within was surely our first great sin against the living Earth. It has haunted us ever since and combustion could now be our *auto da fé*, and the cause of our extinction.

Beyond the Terminus

Like the Norns in Wagner's *Der Ring des Nibe-lungen*, we are at the end of our tether, and the rope, whose weave defines our fate, is about to break.

Gaia, the living Earth, is old and not as strong as she was two billion years ago. She struggles to keep the Earth cool enough for her myriad forms of life against the ineluctable increase of the sun's heat. But to add to her difficulties, one of those forms of life, humans, disputatious tribal animals with dreams of conquest even of other planets, has tried to rule the Earth for their own benefit alone. With breathtaking insolence they have taken the stores of carbon that Gaia buried to keep oxygen at its proper level and burnt them. In so doing they have usurped Gaia's authority and thwarted her obligation to keep the planet fit for life; they thought only of their own comfort and convenience.

Some time towards the end of the 1960s I walked along the quiet back lane of Bowerchalke village with my friend and near neighbour William Golding; we were talking about a recent visit I had made to the Jet Propulsion Laboratory in California and the idea of searching for life on other planets. I told him why I thought that both Mars and Venus were lifeless and that the Earth was more than just a planet with life, and why I saw it somehow in certain ways alive. He immediately said, 'If you intend to put forward so large an idea you must give it a proper name, and I suggest that you call it Gaia.' I was truly grateful to have his gift of this simple, powerful name for my ideas about the Earth. I gladly accepted it then as a scientist acknowledging an earlier literary reference, just as others in previous centuries referred to Gaia when naming the Earth sciences geology, geography and so on. At that time I knew little of Gaia's biography as a Greek goddess and never imagined that the New Age, then just beginning, would take Gaia as a mythic goddess again. In a way, however harmful this has been to the acceptance of the theory in science, the New Agers were more prescient than the scientists. We now see that the great Earth system, Gaia, behaves like the other mythic goddesses, Khali and Nemesis; she acts as a mother who is nurturing but ruthlessly cruel towards transgressors, even when they are her progeny.

I know that to personalize the Earth System as Gaia, as I have often done and continue to do in this book, irritates the scientifically correct, but I am unrepentant because metaphors are more than ever needed for a widespread comprehension of the true nature of the Earth and an understanding of the lethal dangers that lie ahead.

After fifty years living with the concept of Gaia I thought I knew her, but I realize now that I underestimated the severity of her discipline. I knew that our self-regulating Earth had evolved from those organisms that left a better environment for their progeny and by the elimination of those who fouled their habitat, but I never realized just how destructive we were, or that we had so grievously damaged the Earth that Gaia now threatens us with the ultimate punishment of extinction.

I am not a pessimist and have always imagined that good in the end would prevail. When our Astronomer Royal, Lord Martin Rees, former President of the Royal Society, published in 2004 his book *Our Final Century*, he dared to think and write about the end of civilization and the human race. I enjoyed it as a good read, full of wisdom, but took it as no more than a speculation among friends and nothing to lose sleep over.

I was so wrong; it was prescient, for now the evidence coming in from the watchers around the world brings news of an imminent shift in our climate towards one that could easily be described as Hell: so hot, so deadly that only a handful of the teeming billions now alive will survive. We have made this appalling mess of the planet and mostly with rampant liberal good intentions. Even now, when the bell has started tolling to mark our ending, we still talk of sustainable development and renewable energy as if these feeble offerings would be accepted by Gaia as an appropriate and affordable sacrifice.

We are like a careless and thoughtless family member whose presence is destructive and who seems to think that an apology is enough. We are part of the Gaian family, and valued as such, but until we stop acting as if human welfare was all that mattered, and was the excuse for our bad behaviour, all talk of further development of any kind is unacceptable.

So often when disaster visits we still cry, 'How could God have let this happen?' And now that there is a probability that most of us will perish, can belief in God continue? Darwin once described the evolutionary process as 'clumsy, wasteful, blundering, low and horribly cruel'. But surely not as cruel, or as culpable, as we have been and still are to the rest of life on Earth; especially since so many other innocent organisms will share our fate.

It would be easy to think of ourselves and our families as incarcerated in a planet-sized condemned cell – a cosmic death row – awaiting inevitable execution. The days and years will pass, the seasons continue and we will be fed and entertained, and if we have faith we will ask God for a reprieve. Some like Sandy and me will probably cheat the executioner and die before our time is due; the cruel consequences will come for our children and grandchildren.

I am a scientist and think in terms of probabilities not certainties and so I am an agnostic. But there is a deep need in all of us for trust in something larger than ourselves, and I put my trust in Gaia, and declared it in my autobiography, *Homage to Gaia,* in 2000. Was ever a trust so severely tested?

In certain ways the human world is re-enacting the tragedy of Napoleon's advance on Moscow in 1812. In September of that year, when he reached the Russian capital, he had already gone too far, and his precious supplies were daily being consumed while he consolidated his capture. He was unaware that the irresistible forces commanded by General Winter were siding with the Russians, allowing them to counter-attack and regain their losses. The only way he could have avoided defeat was an immediate and professionally executed retreat so that his army could remain intact to fight another time. The quality of generalship is measured in military circles by the ability to carry through and organize a successful retreat.

The British remember with pride the successful withdrawal of their army from Dunkirk in 1940, and do not see it as an ignominious defeat. It was certainly not a victory, but it was a successful and sustainable retreat. The time has come when all of us must plan a retreat from the unsustainable place that we have now reached through the inappropriate use of technology; far better to withdraw now while we still have the energy and the time. Like Napoleon in Moscow we have too many mouths to feed and resources that diminish daily while we make up our minds. The retreat from Dunkirk was not just good generalship: it was aided by an amazing expression of spontaneous unselfish good will from those numerous civilians who willingly risked their lives and their small boats to cross the Channel to rescue their army. We need the people of the world to sense the real and present danger so that they will spontaneously mobilize and unstintingly bring about an orderly and sustainable withdrawal to a world where we try to live in harmony with Gaia.

Economists and politicians have to square the utter necessity of a rapid and controlled shutdown of emissions from fossil-fuel burning with the human needs of civilization. Economic growth is as addictive to the body politic as is heroin to one of us; perhaps we have to keep the craving in check by using a safer substitute, an economist's methadone. I would suggest again that the mobile phone, the internet and entertainment from computers are moves in the right direction; they use time and energy that might otherwise be spent travelling by car or aircraft. Moreover, there is information technology and the efficient use of energy, for example using the ultra efficient white light emitting diodes (WLEDs) to see at night. Should technology of this kind become the main source of economic growth it would let us spend our lives harmlessly and fill some of the time that now we use in fuel-consuming travel. To an extent we are evolving that way.

Until quite recently, although many of us were aware that serious environmental change could happen and believed the predictions of the IPCC, somehow our knowledge seemed theoretical and academic, not indicating that something deadly was imminent. It was a small event that awakened me to these dangers. Fear crystallized as sharp needles in the supersaturated spaces of my mind when, in October 2003, my near neighbours, Christine and Peter Hadden, told me of plans to erect giant wind turbines in the countryside near our homes. Suddenly I realized what our politicians meant by sustainable development and renewable energy, and what it would do to the last remaining good countryside of West Devon. I could almost hear them say, 'Let us harvest the wind for energy, and plant bio fuel crops to keep the cars of urban voters running. We can do it without polluting the air or tangling with that nasty, dirty, fearful nuclear stuff.'

By good countryside I mean farming land and communities that live well with the Earth and represent an ecosystem which, although dominated by people, has ample room left for woodlands, hedgerows and meadows. Most of southern England was like this before 1940, and the largest remaining parts are in the West Country, especially in Devon. In my mind these last remaining areas of countryside were the face of Gaia, and it was about to be sacrificed. It was this that awakened my fury, and made me fully aware of the coming crisis of global heating. To make good countryside into industrial parks for wind energy merely as a gesture to prove their environmental credentials showed how far our leaders were from understanding our peril. To keep their urban enclaves comfortable, they would devastate by industrial development the remaining areas of good countryside.

I moved to West Devon forty-four years ago to escape the bulldozers that were destroying the Wiltshire hedgerows and meadows. Unwisely I thought that the gentle farmland of Devon was too poor to be developed and would let me live out my life in a countryside I loved. I had not allowed for incessant ideological good intentions and the near-religious belief in renewable energy and sustainable development for the good of us all.

They call Sandy and me 'NIMBYs' because we fight their final solution to the energy problem. Perhaps we are NIMBYs, but we see those urban politicians as like some unthinking physicians who have forgotten their Hippocratic Oath and are trying to keep alive a dying civilization by useless and inappropriate chemotherapy when there is no hope of cure and the treatment renders the last stages of life unbearable.

But I was just as mistaken as the urban greens. I had forgotten that the engineers could use a device like a wind turbine sensibly. They designed gigantic turbines that collected and delivered wind energy from places far out in the ocean. There is no need to devastate what is left of Britain's unique and beautiful countryside. Indeed, ocean turbines deliver more power efficiently.

So is our civilization doomed, and will this century mark its end with a massive decline in population, leaving impoverished few survivors in a torrid society ruled by warlords on a hostile and disabled planet? I hope that it will not be that bad; once a technically advanced nation wakes up to its responsibility, perhaps in response to our alarm call, they will say 'we can fix it.' They might use something like space-mounted sunshades or Latham's floating nuclei generators that put white reflecting clouds across the ocean surface. Technological fix it may be, but if it works we have only ourselves to blame if we do not take advantage.

Sunshades for cooling the Earth are more valuable than they might at first appear; they could wholly neutralize the harmful effects of unscheduled methane releases. They might even provide an adjustable remedy ready to off-set the global heating should the methane clathrates of the ocean suddenly escape into the atmosphere. Keeping in mind the similarity of the Earth's physiology to that of a human, it is useful to compare such a technological fix with the use by paramedics of oxygen for heart failure and breathing difficulty, or a pressure pad for haemorrhage – something temporary, to keep a patient alive until they reach the full services of a hospital.

By itself this fix will do no more than buy us time to change our damaging way of life, because if we continue to burn fossil fuels and let the carbon dioxide rise in abundance, ocean life, essential to the health of Gaia, will be further damaged. But we may risk it because time is needed to install equipment for carbon sequestration and for nuclear fusion and whatever forms of economically sensible renewable energy become available. In the longer term we have to understand that however benign a technological solution may seem it has the potential to set humanity on a path to the ultimate form of slavery. The more we meddle with the Earth's composition and try to fix its climate, the more we take on the responsibility for keeping the Earth a fit place for life, until eventually our whole lives may be spent in drudgery doing the tasks that previously Gaia had freely done for over three billion years. This would be the worst of fates for us and reduce us to a truly miserable state, where we were forever wondering whether anyone, any nation or any international body could be trusted to regulate the climate and the atmospheric composition. The idea that humans are yet intelligent enough to serve as stewards of the Earth is among the most hubristic ever.

So what should a sensible European government be doing now? I think we have little option but to prepare for the worst and assume that we have already passed the threshold. Like paramedics, their first priority is to keep the patient, civilization, alive during the journey to a world that at least is no longer undergoing rapid change. We face unrestrained heat, and its consequences might be with us within no more than a few decades. We should now be preparing for a rise of sea level, spells of near intolerable heat like that in Central Europe in 2003, and storms of unprecedented severity. We should also be prepared for surprises, deadly local or regional events that are wholly unpredictable. The immediate need is secure and safe sources of energy to keep the lights of civilization burning and for the preparation of our defences again the rising sea level.

The foolish mistake that I frequently make is to take the statements of green pundits seriously. The green movement has done immense harm and cost us dearly. They have unconsciously become the mouthpiece of the carbo fuel industries. They have done so by telling lies about the nuclear industry. Thus, there is a general belief that the word Fukushima refers to a great nuclear fission catastrophe, one that killed over 20,000 people. What in fact happened at Fukushima was a Tsunami. True, a nuclear fission power station was damaged, but no one was killed. The deaths were the consequence of the tidal wave that devastated the Japanese district of Fukushima.

As I mentioned earlier, intelligent engineers soon realised that there was no need to site wind turbines on the English countryside. They operated far more efficiently way out on the ocean surface, where the wind was stronger and more constant. It seems that a considerable proportion of the power used by the United Kingdom will come from a combination of nuclear fission power stations and renewable energy from places where it can be efficiently harvested.

In several ways we are unintentionally at war with Gaia, and to survive with our civilization intact we urgently need to make a just peace with Gaia while we are strong enough to negotiate and not a defeated, broken rabble on the way to extinction. We may need restrictions, rationing and the call to service that were familiar in wartime and in addition suffer for a while a loss of freedom. But as the climate worsens individual nations will need more and more to address disasters locally as they happen. In a sense, the great party of the twentieth century, with its extravagant overspending and its war games, is over. Now is the time for washing up and throwing out the debris.

I ask Earth scientists who so dislike my image of a living Earth, to consider metaphor seriously as a path to the primitive feelings of the unconscious part of our minds. We are two sexes who respond differently and both metaphors may be needed. We belong to the family of Gaia and are like a revolting teenager, intelligent and with great potential, but far too greedy and selfish for our own good.

Men and women both need to be aware of what we are missing. Already for most of us the artificial world of the city is the whole of our lives and we think that to survive all we need is to be streetwise. But even in the city a few remnants of the natural world continue in the parks and gardens. Make the most of them, for they continue to die away, as does the countryside many know and love; they are precious indeed.

Few of us now can change our lives sufficiently to express our allegiance to Gaia, but I suspect the changes soon to come will force the pace, and just as civilization ultimately benefited in the earlier dark ages from the example of those with faith in God, so we might benefit from those brave ecologists with trust in Gaia. The monasteries carried through that earlier dark age the hard-won knowledge of the Greek and Roman civilizations, and perhaps these present-day guardians could do the same for us. Despite all our efforts to retreat sustainably, we may be unable to prevent a global decline into a chaotic world ruled by brutal warlords on a devastated Earth. If this happens, we should think of those small groups of monks in mountain fastnesses like Montserrat or on islands like Iona and Lindisfarne who served this vital purpose.

Few travellers from the north would go to the tropical south without antimalarial drugs, or to the Middle East without checking how the local war was progressing. By comparison our journey into the future is amazingly unprepared.

Where people know well the local danger, as in Tokyo, they prepare for the earthquake to come. When the threats are global in scale we ignore them. Volcanoes, like Tamboura, Indonesia, in 1814 and Laki, Iceland, in 1783, were much more powerful than was Pinatubo in the Philippines (1991), or Krakatoa in Indonesia (1887). They affected the climate enough to cause famine, even when our numbers were only a tenth of what they are now. Should one of these volcanoes stage a repeat performance, do we have now enough stored food for tomorrow's multitudes? If part of the Greenland or Southern glaciers slid into the sea, the level of the sea might rise by a metre all over the world. This event would render homeless millions of those living in coastal cities. Citizens would suddenly become refugees. Do we have the food and shelter needed when cities such as London, Kolkata, Miami and Rotterdam become uninhabitable?

We are sensible and we do not agonize over these possible doom scenarios. We prefer to assume that they will not happen in our lifetimes. We take them no more seriously than our forefathers took the prospect of Hell, but the thought of appearing foolish still scares us. An old verse goes, 'They thieve and plot and toil and plod and go to church on Sunday. It's true enough that some fear God but they all fear Mrs Grundy.' In science we have our Drs Grundy also, and they are all too eager to scorn any departure from the perceived dogma. Scientists and science advisers are afraid to admit that sometimes they do not know what will happen. They are cautious about their predictions and do not care to speak in a way that might threaten business as usual. This tendency leaves us unprepared for a catastrophe such as a global event that is wholly unexpected and unpredicted – something like the creation of the ozone hole but much more serious; something that could throw us into a new dark age.

We can neither prepare against all possibilities, nor easily change our ways enough to stop breeding and polluting. Those who believe in the precautionary principle would have us give up, or greatly decrease, burning fossil fuel. They warn that the carbon-dioxide byproduct of this energy source may sooner or later change, or even destabilize, the climate. Most of us know in our hearts that these warnings should be heeded but know not what to do about it. Few of us will reduce their personal use of fossil-fuel energy to warm, or cool, their homes or drive their cars. We suspect that we should not wait to act until there is visible evidence of malign climate change – for by then it might be too late to reverse the changes we have set in motion. We are like the smoker who enjoys a cigarette and imagines giving up smoking when the harm becomes tangible. Most of all we hope for a good life in the immediate future and would rather put aside unpleasant thoughts of doom to come.

We cannot regard the future of the civilized world in the same way as we see our personal futures. It is careless to be cavalier about our own death. It is reckless to think of civilization's end in the same way. Even if a tolerable future is probable it is still unwise to ignore the possibility of disaster.

One thing we can do to lessen the consequences of catastrophe is to write a guidebook for our survivors to help them rebuild civilization without repeating too many of our mistakes. I have long thought that a proper gift for our children and grandchildren is an accurate record of all we know about the present and past environment. Sandy and I enjoy walking on Dartmoor, much of which is featureless moorland. On such a landscape it is easy to get lost when it grows dark and the mists come down. We usually avoid this mishap by making sure that we always know where we are and what path we took. In some ways our journey into the future is like this. We can't see the way ahead or the pitfalls but it would help to know what the state is now and how we got here. It would help to have a guidebook written in clear and simple words that any intelligent person can understand.

No such book exists. For most of us, what we know of the Earth comes from books and television programmes that present either the single-minded view of a specialist or persuasion from a talented lobbyist. We live in adversarial, not thoughtful, times and tend to hear only the arguments of each of the special-interest groups. Even when they know that they are wrong they never admit it. They all fight for the interests of their group while claiming to speak for human-kind. This is fine entertainment, but what use would their words be to the survivors of a future flood or famine? When they read them in a book drawn from the debris would they learn what went wrong and why? What help would they gain from the tract of a green lobby-ist, the press release of a multinational power company, or the report of a governmental com-mittee? To make things worse for our survivors, the objective view of science is nearly incompre-hensible. Scientific papers and books are so arcane that scientists can only understand those of their own speciality. I doubt if there is anyone, apart from these specialists, who can under-stand more than a few of the papers published in *Science* or *Nature* every week.

Scan the shelves of a bookshop or a public library for a book that clearly explains the pres-ent condition and how it happened. You will not find it. The books that are there are about the evanescent things of today. Well-written, enter-taining, or informative they may be, but almost all of them are in the current context. They take so much for granted and forget how hard won was the scientific knowledge that gave us the comfortable and safe life we enjoy. We are so ignorant of those individual acts of genius that established civilization that we now give equal place on our bookshelves to the extravagance of astrology, creationism and homeopathy. Books on these subjects at first entertained us or titillated our hypochondria. We now take them seriously and treat them as if they were reporting facts.

Imagine the survivors of a failed civilization. Imagine them trying to cope with a cholera epidemic using knowledge gathered from a tattered book on alternative medicine. Yet in the debris such a book would be more likely to have survived and be readable than a medical text.

What we need is a book of knowledge written so well as to constitute literature in its own right. Something for anyone interested in the state of the Earth and of us – a manual for living well and for survival. The quality of its writing must be such that it would serve for pleasure, for devotional reading, as a source of facts and even as a primary school text. It would range from simple things such as how to light a fire, to our place in the solar system and the universe. It would be a primer of philosophy and science – it would provide a top-down look at the Earth and us. It would explain the natural selection of all living things, and give the key facts of medicine, including the circulation of the blood, the role of the organs. The discovery that bacteria and viruses caused infectious diseases is relatively recent; imagine the consequences if such knowledge was lost. In its time the Bible set the constraints for behaviour and for health. We need a new book like the Bible that would serve in the same way but acknowledge science. It would explain properties like temperature, the meaning of their scales of measurement and how to measure them. It would list the periodic table of the elements. It would give an account of the air, the rocks, and the oceans. It would give schoolchildren of today a proper understanding of our civilization and of the planet it occupies. It would inform them at an age when their minds were most receptive and give them facts they would remember for a lifetime. It would also be the survival manual for our successors. A book that was readily available should disaster happen. It would help bring science back as part of our culture and be an inheritance. Whatever else may be wrong with science, it still provides the best explanation we have of the material world.

It is no use even thinking of presenting such a book using magnetic or optical media, or indeed any kind of medium that needs a computer and electricity to read it. Words stored in such a form are as fleeting as the chatter of the internet and would never survive a catastrophe. Not only is the storage media itself short-lived but its reading depends upon specific hardware and software. In this technology, rapid obsolescence is usual. Modern media is less reliable for long-term storage than is the spoken word. It needs the support of a high technology that we cannot take for granted. What we need is a book written on durable paper with long-lasting print. It must be clear, unbiased, accurate and up to date. Most of all, we need to accept and to believe in it at least as much as we did, and perhaps still do, the World Service of the BBC.

In the dark ages of our earlier history the religious orders in their monasteries carried through the essence of what makes us civilized. Much of this knowledge was in books, and the monks took care of them and read them as part of their discipline. Sadly, we no longer have callings like this. The vast collection of knowledge that is now available is more than any one person could hold. Consequently it is divided and subdivided into subjects. Each subject is the province of professionally employed specialists. Most are expert in their own subject but ignorant of the others – few have a sense of vocation.

Apart from isolated institutes like the National Centre for Atmospheric Research perched on a mountainside in Colorado, there are no equivalents of the monasteries. So who would guard the book? A book of knowledge written with authority and as splendid a read as Tyndale's Bible might need no guardians. It would earn the respect needed to place it in every home, school, library and place of worship. It would then be to hand whatever happened.

Meanwhile in the hot arid world survivors gather for the journey to the new Arctic centres of civilization; I see them in the desert as the dawn breaks and the sun throws its piercing gaze across the horizon at the camp. The cool fresh night air lingers for a while and then, like smoke, dissipates as the heat takes charge. Their camel wakes, blinks and slowly rises on her haunches. The few remaining members of the tribe mount. She belches, and sets off on the long, unbearably hot journey to the next oasis.

Further Reading

The French savant, Bruno Latour's book, *Facing Gaia* (2017) is, I think, essential reading for those who wish to grasp the essence of Gaia theory. On reading it recently, I felt I was seeing for the first time the concept that has occupied me throughout the last fifty-five years. Self-regulation by feedback systems is notoriously difficult to explain in speech or writing. Latour comes closer to this goal than any other writer I have yet encountered.